ANNA'S STORY

ANNA'S STORY

In Pursuit of the Mysterious Missing Day

WRITTEN AND ILLUSTRATED BY
JEFFERY L. THOMPSON

Author and Illustrator

Publishing Futures

Copyright © 2021 by Written and Illustrated by Jeffery L. Thompson

All rights reserved. No part of this book may be reproduced in any manner whatsoever without written permission except in the case of brief quotations embodied in critical articles and reviews.

First Printing, 2021

Dedication
To Beverly from Jeffery

I wanted very much to learn to draw, for a reason that I kept to myself: I wanted to convey an emotion I have about the beauty of the world.
Richard P. Feynman

Contents

Dedication   v

1. ANNA   1
2. MYSTERY   5
3. NICOLAUS   15
4. THE LITTLE COMMENTARY   27
5. THE PATHS OF THE PLANETS   37
6. REVOLUTION   50
7. PHILOSOPHY AND THEOLOGY   56
8. EPILOGUE   64
9. Science, Technology, Engineering and Math   68

I

ANNA

Anna Schilling was the most unusual of ladies in any year, including the current one, but extraordinarily singular in 1529. She lived in the city of Danzig, Royal Prussia (now Gdansk, Poland). In addition to being exceptionally beautiful, she was also well schooled and possessing a towering intellect.

Heinrich (Henry) Kruger, a wealthy Dutch expatriate merchant trader, living in Danzig, wanted desperately to have a son who would carry on the family business, but instead, he got Anna, born in 1490. While appearing to be delighted with a healthy child, he must have also thought: "Gott im Himmel (German for God in Heaven), another dowry to finance."

Within a half dozen years, it was glaringly obvious that Anna was more than somewhat precocious at everything, numbers especially. Henry was entirely delighted to humor and encourage her interests, including a birthday gift of a Chinese abacus along with study notes from an Italian abacus school. Ever since Fibonacci and his interesting numbers, circa 1300, and the building of the great dome in Florence by Brunelleschi (1420 to 1436), abacus schools sprouted up all over Italy, particularly in the major trading cities, e.g., Venice, a regular trading destination for the Kruger merchants.

Anna Schilling

Henry Kruger was one of Danzig's first merchants to adopt the new Indian numerals (often called Arabic because of the route they took to Europe), and the infallible calculating machine – the abacus – for all his accounting work.

Kruger vacillated wildly in his feelings as he watched Anna racing towards adolescence and marriage – he hoped. There were many days when Henry opined that if only Anna had been a male, a phenomenally successful career would be all but an absolute certainty. As an early teenager, Anna was infinitely better than the best adult male bookkeeper in all the Kruger operations.

She had trailed around after Henry as he went about his work since she could walk on her own. Following her father meant numerous visits to Danzig's docks where Anna quickly picked up the dialogue which swirled about them as father and daughter forged their way through the crowded streets. In Polish, Dutch, German, Italian, Greek, and Russian, Anna could switch from high-born lady to dock worker in midsentence.

Possessed with a sharp wit and a rapier tongue, Anna stood down to no man when she knew she was correct; in weights, measures, and money exchange, Henry marveled as it seemed she was always right. Not once did he ever find an error in her meticulously penned columns of numbers. With her fingers flying around on her favorite abacus, she could calculate sums, differences (subtraction), products, and ratios faster than one could say the numbers, much less attempt even starting to perform any calculation. Only half in jest, Henry thought that if he could not find a suitable husband for Anna, he could send her off to be cloistered in an abacus school: better that than a nunnery.

Now at nearly forty, Anna mused on how she had ended up in a wagon being bumped, thumped, and tossed about on her way towards Spain, a two-month journey from Danzig. Roads such as those between cities and villages in Europe of 1528 were, for the most part, nothing but ruts in the dirt. Impassable mud sinkholes in wet weather and choking dust bins when it was dry.

What, Anna conjectured, did she think she could find in Spain? How can an entire day simply disappear before the watchful glare of professional ships' officers paid to attend daily observation? In 1519 Magellan undertook a secret mission for the Spanish crown: find a westerly route to the spice islands and claim the islands not in clutches of Portugal to be the property of Spain. Five ships with a company of 265 hearty men set out from Seville on 10 August 1519 to sail westward to the spice islands. On 7 September 1522, a mere eighteen survivors that looked more like half ambulatory skeletons than living men made landfall with a single vessel barely able to stay afloat, arriving at Seville from the east: they had circumnavigated the globe.

A tangle of intrigue, deceit, half-truths, and complete lies required almost half as long to sort out as the voyage itself had taken. One of the surviving "captains" had led a failed mutiny hardly an entire month after departure. One of the five ships did abandon the project and turned back as the tiny fleet finally made it around the southern tip of South America. Magellan himself had died (27 April 1521) in a senseless fight with locals in an archipelago of islands now known as the Philippines.

However, two of the surviving ships' officers had kept meticulous journals, writing an account every day of the three years without fail. The real story of the events was slowly pieced together in the courts: except for one. Everyone on shore in Spain announced the date of return as 7 September 1522; the ship's officers and their journals said it was 6 September 1522. An entire day had simply disappeared into thin air, as water might boil away to vapor, leaving not a trace.

News of the circumnavigation itself flashed from port to port, merchant to merchant, sailor to dock worker to sailor, across Europe in a matter of a few weeks. The longer story of the missing day took longer: years to stitch that bolt of cloth into seaworthy sail.

2

MYSTERY

As the early sixteenth century opened, Heinrich Kruger, then Torun (also known as Thorn), had no trouble finding a suitable husband for Anna. In 1515, even though she was by then at the "advanced age" of 25, she was wed to a fellow Dutch expatriate and wealthy merchant trader, Arendt Van der Schilling of Danzig.

Kruger had done well by his daughter; she lived comfortably by standards of the day when having children without anesthesia, antiseptics, or antibiotics and before sterilization was even a word could be considered an "easy life." Anna did have a household staff, cook, nannies, and then tutors for the children so that she could continue her intellectual pursuits.

Although well married, Anna did not cut her ties and gossip lines with the docks or the workers there; travelers' stories and ships' officers were the best news network as they would remain until the telegraph's invention, still some three centuries in the future. The *Mystery of the Missing Day* fascinated Anna the first moment she heard the flimsiest muddled fragments of the story. While Anna enjoyed what she saw as a comfortable life as the wife of a wealthy merchant and mother of healthy children, she could not ignore that things did not seem to be going so smoothly for many less fortunate working people in Europe as the sixteenth Century started.

Life for everyone but the ruling aristocracies and wealthiest merchants was "brutal, harsh, and short."

The Catholic Church was adamant in dictating to one, and all that life had always been this way and would always be so. The world was immutable, unchanging, formed in its entirety just as it was then, and it would remain so until the end of time. Although the Sun still rose in the east every morning and set in the west every evening, extraordinarily little else seemed to be staying the same. An entire day gone missing just because one was traveling could not be possible. The more Anna tried to learn, the more credible the Missing Day story seemed to become. But such an event should not be possible. In an age where the word "skeptic" did not yet exist, this unexplained situation was a disconcerting experience.

The most prominent fracture in central Europe seemed to start in 1517 with Martin Luther's critique of the Catholic Church's corrupt practices, a rebuke which quickly unleashed a maelstrom of deep-seated frustrations. Luther became the most well-known critic of the Catholic Church but was not the first: notably Peter Waldo, Arnold of Brescia, Girolamo Savonarola, John Wycliffe, and Jan Hus—but only Luther succeeded in sparking a broader, lasting movement. As sentiment for reform turned into a rage for revolution, Luther found that he could neither direct nor control his fanatic followers, and the Catholic Church could not repress them. However, they certainly tried to do so.

From an obscure start in 1517, the Lutherans, or Protestants as they preferred to be called, the Reformation quickly turned into a gigantic storm. The Catholic Church tried but failed to silence Luther at the 1521 Diet of Worms, followed by the 1524 – 1525 Peasant Revolt in Germany. At first, it had been Catholics vs. Reformers (Protestants), but it degenerated into a free-for-all of every faction against all others. In less than four years after the Diet of Worms, the battles of religious fanaticism claimed more than 250,000 lives in Germany alone.

In the fragmented politics of Europe, a village could be Protestant one week, Catholic the next while the town just half a day's walk down the road did the reverse and cast its lot with the opposite side. Whole countries took up sides, sometimes only to switch teams with the next sovereign or splinter apart into warring camps. Astrology became the rage of the times, particularly among the upper classes as they were aghast at the thought of seeing their newly acquired comforts demolished by the mobs that seemed to be afoot everywhere.

The very fabric of the world which the Catholic Church had declared to be immutable seemed to be disintegrating. In such unsettled times, everyone wanted to know what might be their forecast. Since one cannot hold a mirror up to the future to see what might be ahead, casting horoscopes became all the rage.

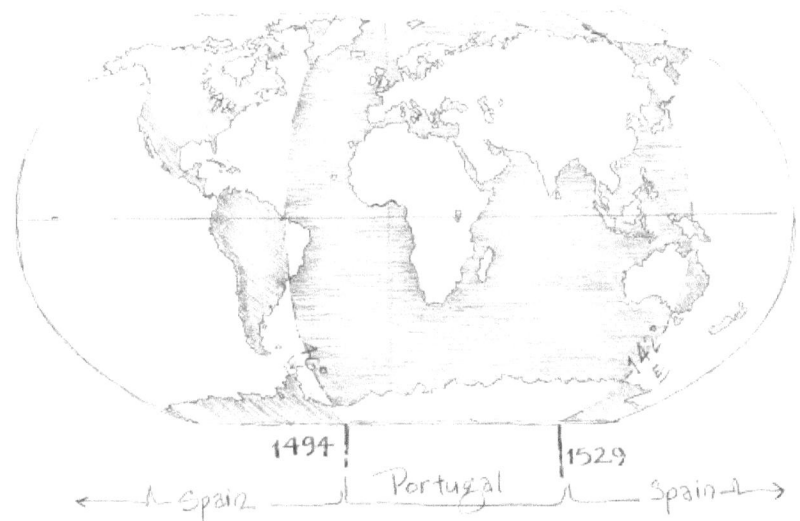

Figure 4: A sketch map of the World (as it's known now) and the lines dividing up the world for Spain and Portugal, first with the 1494 Treaty of Tordesillas. Then after the Magellan Expedition and the realization that there was another ocean, the Pacific, Spain and Portugal drew picked another meridian line at 142 degrees East Longitude. The second line from the Treaty of Zaragoza was roughly the antipode of 42 degrees West Longitude, giving Spain and Portugal approximately equal areas to colonize and exploit.

Anna had become a serious student of astronomy. In the 1500s, astronomy was essentially a synonym for mathematics, geometry in particular; Anna learned Latin to read and study all the available writings on the subjects. As a result of learning languages on the docks of Danzig, it was relatively easy for Anna to also acquire classical Greek for reading early works on astronomy. The line between astronomy and astrology was fuzzy at best, and Anna humored many in her social circle by using her astronomy skills to play at writing horoscopes.

When there is a disagreement in an account, Anna reasoned, it can only be that the two sides are either using different information or different computation methods. If everyone uses the same numbers and performs the same calculations, the results will be identical. Yet, in the *Mystery of the Missing Day*, it appeared to be increasingly necessary to embrace the impossible: same numbers and the same addition methods produced different sums as to how many days were between 10 August 1519 and 6 September 1522.

A school child should be able to do this, and still, professional adults were getting two different answers. It is as if one were sitting at dinner with a friend and looking at the plate, only to have one's companion remark that there is no plate on the table at all. There is a plate, or there is no plate; it is not a matter of opinion.

It is a testable, verifiable fact.

Anna went to great lengths to collect all the best astronomy instruments of the day, mainly from Nuremberg watchmakers turned astronomical instrument builders. Anna built sundials and collected her own data, read everything she could find in print on astronomy, and still, she could not solve the mystery of the missing day. The more Anna studied, the more convinced she became that there must be an error in the ships' logs that recorded the passing of each day on the Magellan expedition.

In 1494 following the voyages of Columbus, Spain and Portugal finally agreed upon a plan to divvy up the Atlantic Ocean's spoils and its shores: The Treaty of Tordesillas, signed by Spain on 2 July 1494 and

Portugal on 5 September 1494. With the Pacific's discovery, a whole new ocean that contained a plethora of islands just ripe for conquest by the rapacious required a new plan of allocation to divvy up the goods. Thus, Spain and Portugal agreed to meet in Zaragoza (or Saragossa in English) located in Aragon's independent state in Spain to hammer out a new (more encompassing) deal to be signed in the spring of 1529.

Spain and Portugal were so arrogant that they did not bother to include anyone else when they parceled out ownership of nearly half the globe: not even the Pope was invited to this party. None of the other rulers in Europe and certainly not any of the people living on the appropriated lands had a voice in the decision-making.

After the Magellan Expedition, the realization that there was another ocean, the Pacific, Spain, and Portugal drew another meridian line at 142 degrees East Longitude. The second line from the Treaty of Zaragoza was roughly the antipode of 42 degrees West Longitude, giving Spain and Portugal approximately equal areas to colonize and exploit.

Through her network of information and gossip, Anna learned in late 1527 of the treaty negotiations, which were to be held at Zaragoza. Verified copies of the ships' logs from the voyage of Magellan were to be there as part of the reference documents available to the negotiators. Now Anna was on a mission. intersect with the Magellan log copies and review the accounts for herself. This mission presented no small task. Just getting from Danzig (Gdansk) to Spain in 1528 was a major enterprise (at least two months of hard travel overland), never mind trying to see the ships' logs.

Anna poured over the Kruger company and Schilling operation accounts, looking for shipments bound for Spain. Both companies dealt primarily with trading metals, not gold and silver. Theirs were far more pedestrian materials, including bronze, brass, and cast iron, the stuff of which canon barrels were made. In the 1500's it was a brisk business as canon barrels could be worth their weight in gold when negotiating the balance of power between nations. With her father and husband's friends and business contacts, Anna had many connections in extremely high places. It was not too difficult for her to charm her way into a scheduled trading expedition party heading for Spain.

Technically, Anna was still married to Arendt Schilling, but he had grown tired of a woman who was better read, more worldly, had traveled more widely, and was the source of an endless torrent of questions about how things worked. Anna seemingly queried him about everything, from the trivial such as the hoist on a ship, or its bilge pump to the profound, for example, the sun's temperature and how far away it was. Every question led to ten more. It just never stopped. Even in the bed-chamber after the last candle of the evening had sputtered out, she once sat at the window and inquired, "look at the stars. Why do they seem to twinkle? If you could travel to one, would it be hot or cold?" Poor Arendt got a headache every time he contemplated talking with his wife – as if he could answer even one of her questions.

Finally, he had had enough and made a business arrangement for a separation allowing Anna to keep their house, her dowry, and maintain her manner of living while he found lodgings elsewhere. He also found a significantly younger, beautiful, and – one suspects – much less scholarly lass.

The first overland traders' caravan to Spain left Danzig in March, and Anna was included in the company's administrative contingent as a translator and interpreter. The caravans of the larger, well-financed trading companies traveled as self-contained entities. They brought tents and bedrolls along with kitchen wagons which provided the morning and evening meals. The kitchen wagons and their attending guards played a sort of leapfrog pattern. One wagon would travel ahead, buying fresh foodstuffs along the way as the other kept up with travelers' main body, then as the fresh items were consumed, the wagons would switch roles. Inns, such as they were, along the road were notorious as meeting places for highwaymen to scout their next victims (anyone foolish enough to travel without armed escorts). A well-equipped caravan under the management of experienced commanders could ordinarily travel 20 to 40 miles per day, depending on the weather – which they did, continuously, seven days a week. Every day was the same routine, up before dawn, a quick meal, packed and ready to travel by first light and 10 to 14 hours of nonstop walking or riding on a horse or in a wagon depending on one's station in the hierarchy of the caravan.

As the wife of a company owner, Anna was accorded her choice of riding horseback or in a wagon. Either way, she could spend most of the day reading and keeping her own counsel. Anna did not expect to be catered to because of her status. She packed her own kit and carried her own bags, which endeared her to the caravan's hard-working crew.

There were no "rest stops" or convenience stores along the road, so Anna, like all the travelers, would carry a few items of fruit or vegetables to eat during the day. As there were no "facilities" along the way to go to the toilet, one simply stopped, and for the sake of modesty, stepped behind a tree if one were available.

The roads were worn tracks across the countryside, "over hill and dale and through the woods." When it rained, the wagons often became mired down in the mud. In such cases, everyone—including Anna —

pushed or pulled on ropes to free the wagons and keep them moving. Long-distance travel in 1528 was expensive, dangerous, slow, and vacillated alternately between boring and exhausting. Boring was good in that it meant the caravan managers knew the route, avoided trouble, marshaled the guards and Teamsters, as well as keeping the kitchen wagons reasonably well-stocked.

Figure 5. Anna did not like to be catered to so she packed and carried her own bags which endeared her to the hard-working crew of the caravan.

Contemplating five or six months of travel – first to Spain, finding a way to examine the Magellan ships' logs, and then the return to Royal Prussia (Poland) – she carefully selected a few books to read on her travels. The first and number one choice on Anna's list was not even a published book, but rather a hand-copied, anonymous long "letter" of some three dozen pages provided by one of her small circles of mathematically inclined friends. The document came to be known as De Commentariolus, which even in the early years after 1514 when it started circulating, everyone who copied it knew Mikolaj Kopernik had written it, or in English, The Little Commentary by Nicolaus Copernicus. The document was named by Tycho Brahe but not until nearly thirty years after Copernicus died, circa 1571, when Brahe was 25 years old.

There is a well-known invisible university in engineering and science: word of mouth among people of similar interests. The term "invisible college" was coined by Boyle in 1645 to describe meetings of the Philosophical Society. The idea goes by various names in different countries, but it is the root of the cliché among modern-day engineers and scientists that the most productive part of attending a technical conference may be the excursion trip bus rides. Here informal ideas and contacts with similarly interested people are freely exchanged without fear of rebuke or criticism.

Anna was well connected through her husband's merchant trading business, a trade that included books and iron, bronze, salt, and salted fish. As Anna became well-known in the circle of merchant traders as a person keenly interested in astronomy, she heard of the latest books (Frankfurt am Main book fair) and news from the merchants and travelers who arrived via the docks of Danzig.

Through this network of introductions and gossip, Anna knew of Copernicus as an astronomer and mathematician who was a minor canon of the Church at Frauenburg in Warmia (now named as Frombork, Poland), hardly 50 miles from Danzig.

So it was that a well-known acquaintance of Anna passed a handwritten copy of a letter that was becoming increasingly well-read and discussed among the best astronomers of the day. It was written by an anonymous author but unanimously attributed to Copernicus. By what seemed the most fortuitous of coincidences, the Schillings also owned a house in the village Frauenburg overlooking the Vistula Lagoon (or Bay), a pleasant 15-minute stroll top of the hill where the Cathedral was located as well as the Curia of the "anonymous" author.

By 1514 Copernicus was well enough known as an astronomer of some considerable skill that he was invited to attend the conference on calendar reform called by Pope Leo X. Scheduled for autumn and to be held in Rome with a report to be issued by the end of the year, Copernicus declined the invitation. He seemingly ignored it but did submit a report to conference administrator Paul of Middleburg.

As with previous similar efforts on calendar reform, nothing ever transpired from it. However, the fact that Copernicus was invited to the conference was an impressive credential and served to elevate his status. As a child, born in 1473, young Copernicus perhaps liked looking up at the stars as much as any youngster does but being an astronomer was not in the family plan for Nicolaus. Born in Torun (or Thorn) in Royal Prussia, which today is western Poland, Nicolaus was named after his father, a wealthy trader merchant.

Figure 6: University of Krakow

3

NICOLAUS

Torun is on the banks of the nearby river, which was said to be "the golden stream" of Prussia, gold by virtue of the wheat that was shipped downriver to Danzig (Gdansk) and then to most of the Hanseatic cities of Europe. Nicolaus Kopernigk traded primarily in copper, which was sold in Danzig. Young Nicolaus was the youngest of four children, two older sisters, and an older brother. As the custom of the time, it was no doubt expected that Andreas and Nicolaus would follow their father into the family business.

But when Nicolaus was just ten years old, his father died suddenly. Luckily for Nicolaus, his mother, Barbara Watzenrode Copernicus, was from a wealthy patrician family of Torun and her brother Lucas Watzenrode stepped in to provide for young Nicolaus and his brother Andreas after the death of their father. The oldest sister, Barbara, was sent to a Benedictine nunnery, and the younger sister, Katherine, was married off to a prominent Torun businessman and city councilor. But Nicolaus and Andreas were sent to the absolute best schools in Torun. Copernicus attended St. John's School where Lucas Watzenrode had once been a master before continuing his education at the Cathedral School.

At Wloclawek (1488 to 1491), upriver on the Vistula River from Toruń, which prepared pupils for entrance to the University of Krakow, Watzenrode's alma mater in Poland's capital. Nicolaus, together with his teacher, Mikolaj Wódka (Abstemius), built a sundial (in 1489). A restoration of sorts can be seen today at the St. Mary of the Assumption Cathedral Basilica in Torun.

By 1489 Lucas Watzenrode had climbed the Catholic Church's political ladder and was confirmed by Rome as a bishop. The wealth and position of Bishop Lucas Watzenrode as patron and advocate for Nicolaus meant that short of being born into a royal family, young Nicolaus Copernicus was as well connected and supported as one could be. Clearly, Bishop Watzenrode was educating and grooming Nicolaus to follow his footsteps into a position of power and wealth in the Church.

When Copernicus enrolled at the University of Krakow (1491), he did not stay in the student dormitory as almost all students did. Instead, he and Andreas took rooms in the home of a close friend of Bishop Watzenrode, Piotr Wapowski, a wealthy patriarch of Krakow and the assistant to Bishop Freyderyck of Krakow and Gniezno. Here Nicolaus met a son of Piotr Wapowski, Bernard, a contemporary of Nicolaus also enrolled as a student in the University of Krakow and passionately interested in cartography; the two would become close personal friends, a relationship which endured for the rest of their lives.

It would have been easy for Copernicus to have simply coasted from this point on, enjoying a life of privilege, prestige, and power without having to work extremely hard. But Copernicus did just the opposite and quickly distinguished himself as bright, hardworking, and a quick learner with a passion for mathematics.

As Nicolaus was about to enroll for the summer term of 1492, disaster struck: the main building of the University of Krakow, the Collegium Maius, was destroyed by fire. While the university was quick to assure everyone that it would be rebuilt, the reality was that the fire was utterly devastating. The Collegium Maiuse, as the main university building, housed school records, most of the classrooms, and the professors' offices and apartments.

Moreover, in the Copernicus era, the university's library consisted essentially of the faculty's personal books. When one was appointed as a university professor, they were given an apartment but then assigned their personal books to the school. The loss of the Collegium Maius meant that most of the faculty was now homeless, there were few remaining classrooms, the university had lost almost all their books, and all their student records were destroyed.

It was an unmitigated disaster for the institution of the university and personally for Copernicus, who now had seen his world collapse twice in a short span of just ten years: first the death of his father and now the destruction of his intellectual home.

Figure 7 Small detail sketch showing a portion of the ruins of Heidelberg Castle; it opened in 1214 and was partially destroyed in the 1500s. Main part of the castle has been rebuilt several times since the sixteenth century and is a major tourist attraction in the area

Krakow's upper-class citizens took great pride in their university, so the homeless faculty was quickly provided temporary quarters in private homes. Piotr Wapowski was delighted to do his part and have Adalbert of Brudzewo (or Brudzewa), the preeminent scholar of mathematics and astronomy in all of Europe as his invited house guest.

Young Copernicus must have been astonished to realize what had transpired in a mere ten years of his life. Here he was at just nineteen years of age in Krakow, the national capital, at the dinner table of one of the city's most elite families sitting next to the most prestigious math and astronomy professor in the country, if not all of Europe:

Nicolaus Copernicus had come a long way from the village of Torun. Dinner conversation was, of course, almost entirely focused on the state of the university and what was being done to reopen and rebuild the school.

While the university's official position was that they would have the summer term as planned, Adalbert shared private conversations that the reality was that the administration did not even know where more than one-third of the faculty was: were they alive or dead, in town or not? No one knew.

There were few classrooms available in the other two school buildings. The Collegium Maius had housed most of the university's books, of which only a few had survived; in summary, the university was in complete confusion. Adalbert thought it more realistic that it would take months, not weeks, to get organized and resume anything remotely like a normal schedule.

Adalbert had been relatively fortunate in that his office was on the far side of the building from where the fire had started. Thus, he had managed to save a fair portion of his most valuable books and important papers by simply throwing them out of his office window into the waiting arms of students and city folk below: articles, papers, and books that were now with him in the Wapowski residence.

Among the salvaged papers, Adalbert had a collection of observation data on a lunar eclipse which occurred on 12 May 1492 and had been seen in much of Europe. Adalbert had heard of Copernicus being described as bright, clever, and dynamic by other faculty members, so he proposed that Copernicus sign up for courses he was scheduled to teach. With something of a conspiratorial smile, Adalbert said the study for the course would actually be working with him to analyze the eclipse data.

News of the Collegium Maius fire had quickly radiated out from Krakow, reaching Marcin Bylica (also known as Martin Bylica or Marcin Z. Olkusza), one of the most famous alumni of Krakow at the University of Buda by the end of the month. The cities of Buda and Pest are on opposite banks of the Danube River. Bylica immediately sent a collection of the best and most recent astronomical instruments to the University of Krakow to help the school in its rebuilding efforts. Adalbert was now in charge of the new instruments. He wanted to have them ready for the next complete lunar eclipse predicted by many to occur in August; Copernicus seemed like the ideal student to work on such an effort.

A lunar eclipse did occur on 17 August 1492; the new instruments afforded a wealth of new data, which meant Copernicus had a full calendar of study with Adalbert in astronomy and math courses by similar arrangement negotiated by Adalbert with other faculty members. Copernicus was astonished by how new and better instruments produced new and better data, generating new and better ideas and questions. The experience profoundly influenced how Copernicus viewed the relationship of conjecture, theory, data, and resulting opinion.

Nicolaus was still working on the eclipse data and the instruments donated by Bylica when a comet appeared in December and remained visible for the better part of two months. Copernicus spent every clear night making observations of the comet and went from despair in July to elation with a new passion by the Christmas of 1492. The start of a lifelong interest in astronomy is documented by the purchase of many books on the topic. Many of the records and books both still exist.

In the spring of 1493, Adalbert received a long letter and package from Ensisheim Alsace (now France, but then in one of the many independent German states). For the first time in recorded history, a meteorite had been observed on its impact in a wheat field just outside the walls of Ensisheim on the southeast side closest to Battenheim. Souvenir hunters almost immediately began chipping away at the chondrite stone, estimated at an initial weight of 280 lbs. Fortunately, the town Magistrate intervened and had the rock placed in the local cathedral for protection and safe-keeping.

The city and Church sent a long letter, report, and sample to Adalbert as the most well-known astronomer in Europe for his opinion and support of the meteorite's authenticity. Copernicus was at the elbow of Adalbert through all of this, at school and often at the dinner table of the Wapowski residence.

Ensisheim seemed to go into the tourist and souvenir business almost immediately, such that the remaining Ensisheim Meteorite now protected in the city museum weighs only about 120 lbs. However, nearly every museum in the world with a meteorite collection now has a piece of the Ensisheim Meteorite, often called the "King of Meteorites."

Meanwhile, by mid-1493, the University of Krakow was back to near normal functioning with temporary class spaces in many locations. Nicolaus Copernicus appeared to be following the life his uncle Bishop Lucas Watzenrode had laid out for him. But Copernicus had a new passion, one that would consume his every minute outside of Church obligations for the rest of his life: astronomy.

At the end of the 1495 spring term, Nicolaus and Andreas left the University of Krakow without completing their degrees, returned briefly to Torun then Copernicus alone set out for Bologna (Italy) to enroll at the University there in Law. Bernard Wapowski also enrolled in Law at the University of Bologna in parallel with Copernicus; both also joined the "German nation" at the University. While the outside view of Copernicus' actions would still be on the path his uncle Lucas made for him, Nicolaus was on another plan internally. Bologna was an interesting University currently without classrooms or faculty. The students would gather in self-selected "nations" and then hire their own professors, set class schedules, and rent spaces to be used as classrooms.

The university appointed professors as department chairs as part of the administration and worked on infrastructure issues but seemingly did little else.

Figure 8 Sketch of a castle tower in present-day Germany, near what was then Danzig and destroyed in the German Peasants Revolt, one of the first (1524–1525) widespread battles of the Reformation – Counter-Reformation hostilities.

Bologna was something of a crossroad between university and art cities: heading north from Bologna, one can walk to Venice in two days. Another university town, Padua, was also on the road to Venice; or going south, one could be in Florence within one day. Bernard Wapowski seems to have recruited Copernicus to help work on a cartography project of his, a map to include the "new world" of North and South America. Columbus and others' voyages created a demand for entirely new world maps (much more would follow when the Magellan expedition completed its voyage 26 years later).

By 1496, having followed the advised path of uncle (Bishop) Lucas Watzenrode for more than a dozen years as well as riding his coattails, Copernicus realized a significant reward for his patience: he was appointed canon of the Cathedral Chapter of Warmia in Frauenburg, Royal Prussia (now called Frombork, in Poland). The assurance of a canonry in his pocket may have given Copernicus a new sense of security independent of the direct financial support of Bishop Watzenrode. By 1497 Copernicus had obtained rooms with Dominico Maria Di Novara, the department chairman of Astronomy at the University of Bologna, and was acting as an assistant to Di Novara. Although Copernicus was still officially enrolled in law school, his intellectual passions were clearly elsewhere, namely in the sky among the moon and stars.

In the spring of 1500, Copernicus left Krakow, again without completing his intended degree, and trekked off to Rome with tens of thousands of other celebrants for the proclaimed 1500th anniversary of Christianity. Bologna's main road to Rome goes through Florence, then home to Da Vinci, Bellini, Botticelli, with more artistic wood carving shops than butchers, and the site of the great Brunelleschi dome on the Santa Maria del Fiore church. The dome was completed in 1436 with a lantern added (1446), which included an aperture (finished circa 1461). The church floor could also be used as a meridian to more accurately determine the Vernal Equinox.

In Rome, Copernicus could follow the same itinerary precisely used a century earlier by Brunelleschi: studying the Pantheon and the ruins of the basilica of Maxentius to examine in detail the methods used by the Roman engineers more than twelve hundred years before his visit.

While in Rome, Copernicus lectured on mathematics and finally returned to Frauenburg in early 1501. Copernicus was rather skilled at disputation and talked the Bishop of Wamia into allowing Copernicus two more years on his Italian sojourn to study medicine at Padua. Finally, in 1503 Copernicus returned to Frauenburg with a degree in medicine granted at Ferrara (graduation ceremonies were less expensive in Ferrara than Padua).

At this point, Copernicus had spent seven years in Italy, save for his brief return to Frauenburg in 1501, and all together an entire decade at university (1493 to 1503). While Copernicus was at university in Bologna (1496 to 1500), he was only a day's walk to Florence, the home and workplace of Da Vinci, Michelangelo, and Botticelli, among many other artists due to the sponsorship of the Medici family.

Being in Italy's art center captured enough of Copernicus' interests that he learned to paint well enough to make at least one credible self-portrait in oils. By 1500 or soon after, Da Vinci made a study of astronomy, concluding that the ancients, such as Aristarchus, had been correct: the sun was at the center of our universe. The planets all orbited around it. By circa 1505, Da Vinci made extensive notes on the subject in his notebooks. It is not difficult to conjecture that Copernicus, a person deeply interested in astronomy and well-connected at the universities, had heard discussions about Da Vinci's opinions.

Back in Frauenburg, 1503, Copernicus then spent the next four years intellectually thrashing over what he has accumulated over the last fifteen years or so and began writing (some while prior to 1507) a comprehensive view of the solar system (a term invented by Copernicus to describe a heliocentric view of the world and its environs). Writing, revising, editing, rewriting takes another three years.

The meticulous, slow, painstaking pace was born of Copernicus' greatest fear that his work would be summarily dismissed as that of a crank or fool. At last, in 1510, at the age of 37, Copernicus makes a dozen anonymous handwritten copies of his six broadsheet or 40-page manuscript for a highly select group of mathematically astute astronomers and friends.

By the same token, those bright enough to be among the inner circle of Copernicus correspondents are also smart enough to recognize the import of what they read. Thus, in turn, this group makes copies for their trusted, mathematically inclined astronomer friends. No one knows how many copies eventually were circulated but by 1528 when Anna gets her copy, the manuscript had been read and was being discussed as far north as Sweden, west to England and Spain, south to Italy, and east to Russia and Lithuania.

While no copy ever bore the name of the original author, it seemed that just about everyone who read the manuscript knew a minor canon had written it in an obscure corner of Royal Prussia, and his name was Nicolaus Copernicus.

The Little Commentary may have been small by page count. Still, Anna instantly recognized it as monumental in concept: the Sun, not the Earth, was stated to be the center of the known universe, with all the planets save the moon revolving about the Sun, the Earth included. Only the moon was said to revolve around the Earth. Moreover, the earth rotated on a North-South axis which produced the observed day-night phenomena. In his anonymous manuscript, Copernicus presented seven ideas which he termed postulates:

Celestial bodies do not all revolve around a single point.

The center of Earth is the center of the lunar sphere—the orbit of the moon is around the Earth.

All the spheres rotate around the Sun, which is near the center of the Universe.

The distance between Earth and the Sun is an insignificant fraction of Earth and the Sun's distance to the stars, so parallax is not observed in the stars.

The stars are immovable; the daily rotation of Earth causes their apparent daily motion.

Earth is moved in a sphere around the Sun, causing the apparent annual migration of the Sun; Earth has more than one motion.

Earth's orbital motion around the Sun causes the seeming reverse in the direction of the planets' motions.

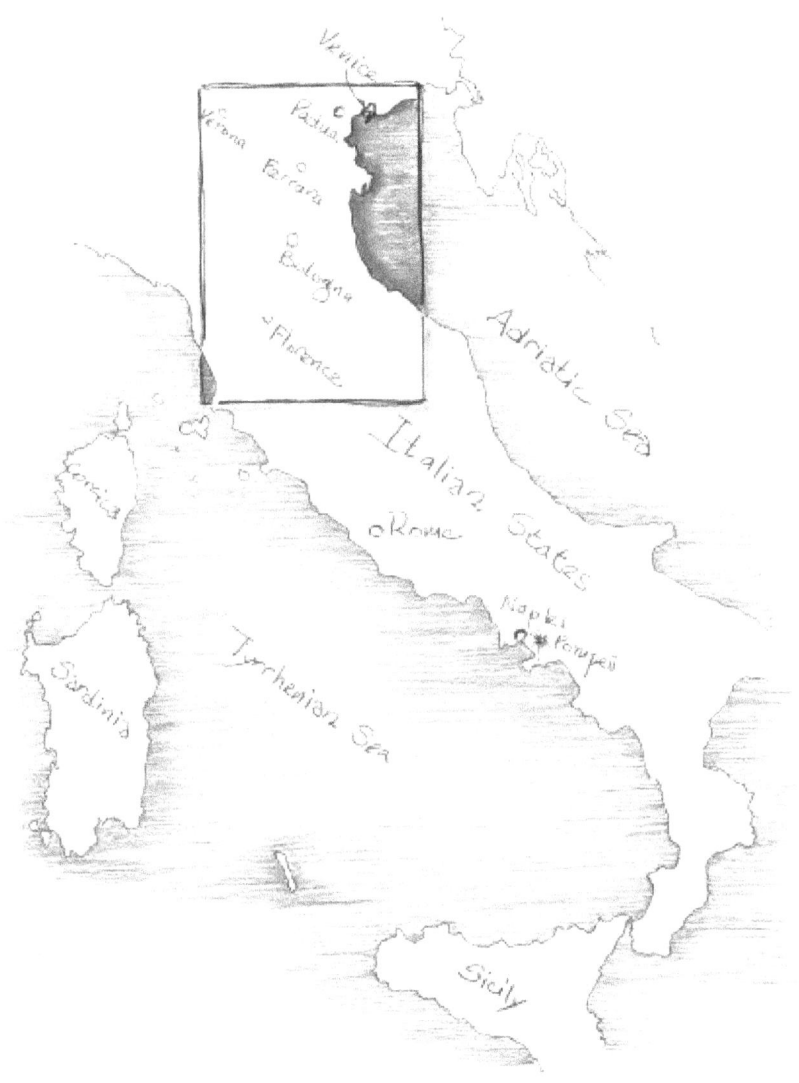

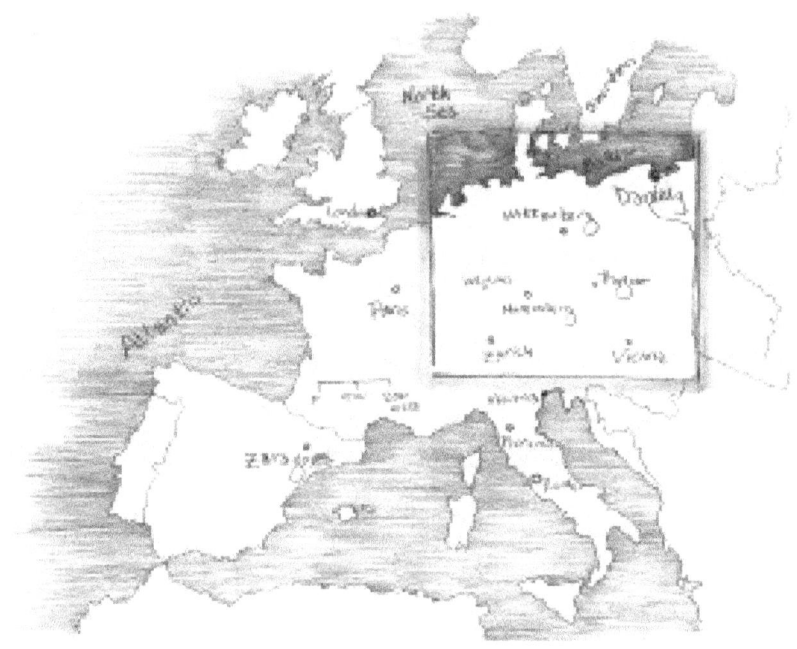

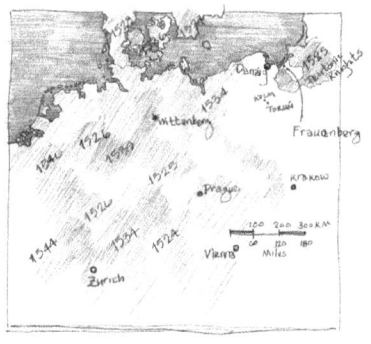

4

THE LITTLE COMMENTARY

Immutability of the Earth as dictated by the Catholic Church had been suffering a series of significant setbacks, starting with the voyages of Columbus (1492 – 1494), which put an entire new continent on the west side of the Atlantic, blocking the direct sailing route to the spices of India. Then the voyage of Magellan (1519 – 1522) put a whole new ocean, the largest on the planet, between the Americas and the Orient. On the theological front, Martin Luther in Germany unleashed a long-stifled torrent of resentment against the Church (starting in 1517), which quickly flashed into open rebellion in many places. One of the first large-scale conflicts became known as the German Peasants' War (1524–1525), where fighting on the eastern fringes of the battle areas spread to within sight of Danzig's western walls. The Catholic Church was not nearly as unassailable in Anna's eyes as it was in the eyes of the Church hierarchy.

The anonymous letter (latter to be titled by Brahe as The Little Commentary) left Anna shocked, fascinated, enthralled and frightened all at once. It was no wonder the document was copied by hand and then only passed carefully and quietly from one trusted friend to another. The Little Commentary was nothing less than incendiary, contradicting Church doctrine on so many points. Brilliant in its audacity, a coherent system view of the universe, and one that did away with the seemingly forced use of clever mathematical inventions to explain the path of the planets. Most puzzling was when the planets seemed to move in one direction, then in reverse, and then again to resume their original direction. All these seemingly engineered constructs to explain the planets' retrograde motion simply vanished with the new arrangement of Sun and planets as postulated by Copernicus. Moreover, this Copernican view suddenly illuminated what we would now call an X-ray vision into a quandary that Anna had found impossibly opaque until now. She could not ingest the Little Commentary fast enough or thoroughly enough to satisfy her questions and curiosity. A two-month journey with nothing much else to do but read was simply perfect. By the time Anna reached Spain, she was certain she had acquired enough of an idea on the workings of the solar system that she could solve the Mystery of the Missing Day. Nothing if not the penultimate salesman Anna convinced the court administrators to allow her to inspect the Magellan ships' logs.

Night and day, she poured over the records, her scratching pen, and clacking abacus droned on and on. The guards assigned to watch her were bored silly and simply spent most of their hours sound asleep. They could not figure her out, she didn't sleep but three or four hours a night, ate very little as if eating were an annoying diversion from her labors, she wore the same clothes for a solid week, hardly pausing long enough to go to the toilet. Yet she was cheerful as a lark. The guards all concluded she was either an agent of the devil, a witch, crazy, or some combination of the three and were delighted almost beyond words when she finally decamped.

The administrators were less enthralled when Anna informed them that both versions of the date calculation are correct. Both the sailors and the people at home in Seville measured time by counting solar days (sunrise, noon, sunset). The ships on the earth's surface were chasing the sun ever westward, at an average rate of about 1 mile per hour (use old Krakow ell units) for three years. Sufficient so that the sailors observed 1122 noon events and the people in Seville saw 1123. Records of the observed lunations (full moons) matched the ships' logs and records from Seville perfectly.

What Anna did not tell the officials is that the reason this was true was due to the earth rotating on its axis while it revolves about the Sun as the ships travel on the surface of (the moving) earth, chasing the sun ever westward, at an average rate of about (1 mile per hour, use Krakow ell units here) for three years. In sidereal terms (measured against the fixed stars), the ships have gained the equivalent of one earth rotation over the stationary people in Seville (measured by local noon sun sighting). Both accounts are correct, and the earth must be in orbit around the sun. It is the only way all the numbers, including those for the moon, will work.

Mystery Solved: but who is going to believe such a story?

At the same instant Anna formulated the question, the obvious answer flashed into her thoughts: The source of her idea in the first place, the person who wrote De Commentariolus, Mikolaj Kopernik, The Little Commentary, Nicolaus Copernicus. She must find a way to meet him.

Meanwhile, in Royal Prussia as well as Poland – death, disease, and pestilence had been grinding away in their usual manner and working overtime to accommodate Anna. Upon returning to Danzig (Gdansk), Anna learned that her husband, Arendt Schilling and Nicolaus Copernicus, had been named joint caretakers for a group of orphans, mutual family cousins both Schilling and Copernicus.

Anna manipulated her inclusion along with Herr Arendt Schilling to the next meeting in Danzig where Copernicus was to be in attendance, a dinner to discuss the necessary support for the orphans. After dinner, Anna found a discrete moment to address Copernicus quietly, "Herr Doktor Kopernick, ich habbe eine frag ge, bitte." Using German, her husband's least favorite language, she asked, "Dr. Copernicus, may I ask a question, please?"

Copernicus was enthralled with the charming lady; Nicolaus was nearly 56, almost twenty years senior to Anna, but he was not unobservant.

Anna then continued in Italian: "do you think it is possible that if I were to remain in this very place and you were to start traveling around the earth, constantly traveling westward chasing the sun, at an average rate of say 40 thousand ell per day that you would see one less noontime sun over the course of 3 years than I will by remaining here?"

More than somewhat incredulously, Copernicus retorted, "where did you get such an idea?"

Anna, with just the slightest hint of a most innocent seraphic smile for which she was known, replied, "from your manuscript; what's more, someone has done it, and they've demonstrated your idea exactly; the Earth does revolve about the sun."

Copernicus stood mute, as motionless as if he were a marble statue. Finally, he stepped to the nearest chair and sat down, looking as though he had been struck a great blow and might collapse. After some time elapsed, he rasped in a whisper barely audible to Anna who was standing right next to him, "Who, where, how?" Anna then quickly recounted the voyage of the Magellan expedition and the *Mystery of the Missing Day*. She included her sojourn to Spain, reading De Commentariolus, examining the ships' logs, and a brief summary of her calculations.

Copernicus stared at Anna as if trying to discern if she were real or merely an apparition that may only be a figment of his imagination. "Who are you? How did you get my writing that was done more than fifteen years ago? It has never been published. I only made a few copies which I selectively distributed only to my closest, most trusted friends."

Anna replied coyly, "In turn, friends make copies for their most trusted friends and I, too, have many friends."

"Apparently," Copernicus responded rather dourly, feeling more than a bit threatened by the realization that his unintended notoriety had grown far beyond his worst nightmare.

In response to Copernicus' first question, Anna gave a summary of her curriculum vitae, barely an outline but providing much more detail than the perfunctory introduction they had before dinner.

Copernicus leaned forward, drawing nearer to Anna, and whispered, "In these times, talk of planets revolving about the sun can get one in a great deal of trouble…, a very great deal."

"Yes, Father Copernicus, so I've observed in my travels. Perhaps I might ask for a private audience where I could confess my unorthodox ideas on that topic."

"I'm not ordained, so I don't hear confessions."

"True enough, but I would imagine you can entertain hypothetical discussions on astronomy." In 1529 the description of an idea as being "hypothetical" served as a code among intellectuals. The official position of the theologian speaking on behalf of the Catholic Church took "hypothetical" to mean the same as "false." This contorted sort of logic would entangle Galileo in the next century, almost exactly a hundred years hence. "Perhaps so, perhaps so." Then observing that the shadows were growing long, Copernicus took his leave to attend Mass, explaining that while he did not say Mass, nonetheless, he was expected to attend.

Now Anna was intent on exploiting the coincidence that the Schillings also owned a house in the tiny village of Frauenburg on the Vistula Lagoon (or Bay); it was their retreat from the crowds and grime of the city and docks of Danzig. Frauenburg was also the site of the Bishop's Castle, and the home of Copernicus – the northwest tower along the outer wall of the Castle. One can find the village on a modern map marked by its Polish name: Frombork.

A week after the meeting with Frau Schilling, a letter was delivered to Father Copernicus inviting him to dinner at the Schilling residence on the next full moon. A common practice of the times as the moon would provide sufficient light to help find your way home after dinner. In 1529 a moonless night was dark beyond anything we can fathom in our modern light-polluted industrial world. Even a light fog rolling in from the Vistula Lagoon, a common event, would obscure the stars so that on a new moon night, it was so dark a standing person could not see their own feet. Attempting to walk any distance at all under such conditions was foolishly perilous.

Anna was elegant at dinner, attired in the latest Italian fashion from Florence and Venice. Father Copernicus traveled to Danzig often enough to recognize current fashion trends as displayed by the wealthier merchant visitors. The resplendent furnishings of the person, the dinner table, and home were noted and impressed the dinner guest. Father Copernicus was as far from being a semiliterate village priest as Anna was from being an impoverished peasant farm laborer.

Dinner conversation ranged over literature, travel, and philosophy with appropriately connected language detours from Polish into German and Italian as well as Latin. Anna had grown up wealthy and had no need for ostentatious display or boastful manners or language; she was brilliant by quiet, comfortably restrained understatement. Copernicus was enthralled, as captivated as a schoolboy of fourteen by both the household's accouterments and the conversation: it had been decades since Copernicus had the opportunity to enjoy a relaxed evening discussing astronomy with a person well versed and highly skilled in the mathematics. Copernicus could hardly recall having such pleasure, which bored through to his very core and resulted in an overwhelming serenity of spirit. At the end of the evening strolling home Copernicus felt as if he could float on the wisps of light fog that drifted ashore from the Vistula Lagoon.

The ceremonial dinner invitation was repeated on the next full moon. Having ascertained some measure of each other, Anna and Mikolaj were more relaxed, and their dinner table conversation turned to astronomy's mathematical technicalities. The glasses and cups set out on the table for dessert became the workings of an impromptu armillary. Both Nicolaus and Anna were sufficiently astute to wait until the butler and cook retired to the kitchen before they allowed their conversation to address the real topic of the evening: moving planets. They switched to Latin or Greek, certain the household staff could not understand a word of their whispered conversation even if they were purposefully eavesdropping.

At a time, still eighty years before the first primitive telescopes would turn to the skies, a century before Rene Descartes invented his Cartesian system along with graphing so one could describe positions by XYZ coordinates, and 150 years before the invention of calculus, the mathematics of astronomy which Anna and Nicolaus discussed were made almost impossibly difficult by the lack of mathematical tools. It was as if today you and a friend were trying to design a modern car engine, but steel has not yet been invented.

Figure 11

Even if you knew how the machine worked, you would find the task of building it or describing it to be so daunting that any rational person would probably abandon the pursuit.

Nicolaus took a calculated risk of making a social blunder that might get him uninvited to any future dinner by inquiring of Anna if he could bring his friend, Fr. Tiedemann Giese, to join them their next dinner. Anna assented immediately, while she did not personally know Tiedemann, he was from the patrician Giese (also spelled Gisze) family of Danzig. Georg Giese, younger brother to Tiedemann, was a prominent merchant trader in the Hanseatic League so the family was known to be of the highest repute with excellent connections known to Anna through both her father and husband.

Tiedemann was seven years the junior of Copernicus but the two found a mutually shared interest in astronomy when in 1516, they worked together as minor canons to write a letter to King Sigismund I the Old of Poland. The correspondence was one to beg for the promised help to repel the brutal assaults of the Order of the Teutonic Knights on neighboring Warmia (where Frauenburg is located). Little surprise in the fact that the long-promised help never materialized. However, Nicolaus and Tiedemann's friendship flourished and endured for the next twenty-seven years, until the death of Copernicus in 1543.

While Copernicus never seems to have been driven to pursue the politics necessary in advancing one's career, Tiedemann Giese was most adroit at scaling the Church's hierarchy (he would become Bishop in 1536). If not entirely content and perhaps privately frustrated, Copernicus appeared resolved to live in his tower inside the castle and outdoor curia, studying astronomy while the world stormed outside.

Even though Copernicus and Giese were distant relatives via some "shirttail cousins", it was their shared interest in astronomy, which brought the two together. They became lifelong friends and confidants.

Tiedemann gained a wide reputation for advocating and publishing letters calling for restraint and tolerance of others' views in theology. During the vicious and lethal religious fanaticism sweeping through Europe in the 1520s and later, Tiedemann maintained an astonishing circle of correspondents, including Phillip Melanchthon, the first systematic theologian of the Protestant Reformation and confidant of Martin Luther. Considering that Giese was a Bishop in the Catholic Church, this was not an inconsequential accomplishment. The well-intended efforts of Tiedemann and his plea fell on deaf ears.

Figure 12

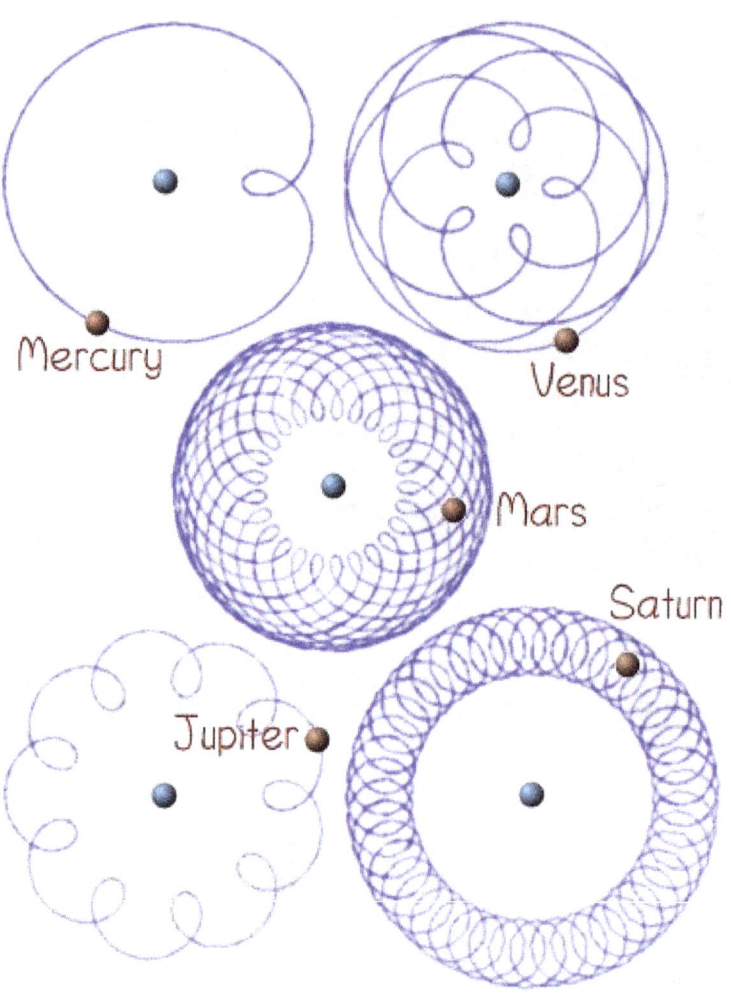

Figure 13

5

THE PATHS OF THE PLANETS

For a serious aspiring student, as Anna was in astronomy and mathematics, there is no other pleasure the equal of working with a great master of one's craft.

Had Anna been an aspiring artist, her experience with Copernicus was the equivalent of working in the studio next to her other contemporaries: Leonardo Da Vinci, Michelangelo (both working in Florence at the time) and Hans Holbein the Younger (then in London).

Anna was probably as infatuated with her situation in astronomy as an ingénue is with her first kiss, and understandably so. While Anna may have been decades beyond her days as an ingénue, she was more than bright enough to recognize the genius of Copernicus.

The experience of student apprentice with master craftsman is a reciprocal one to a very real extent. Today, five centuries after Anna and Nicolaus talked about the mathematics of orbiting planets, modern students and teachers are debating the existence of mirror neurons: a specific category of neurons and neural synapses. The rewarding experience for a master craftsman as a teacher is very real. It is particularly satisfying to a craftsman teacher to have a serious, aspiring student who is also exceptionally bright and skilled. Such students make progress at an extraordinary rate, thus rewarding the teacher with an immediate and satisfying experience in return for their investment of time and effort.

This working relationship is entirely different from the stereotypical fawning graduate student smitten with romantic illusions about her famous professor. These are essentially equal intellects who differ in length of training and depth of professional experience. The student is just as likely to instruct the master by posing an insightful question as the master is to instruct the student by relating experienced wisdom acquired through years of investigation. It was in this context that Anna and Mikolaj began having quasiregular dinners.

Whenever his schedule would allow it, these evenings included Fr. Tiedemann Giese. Tiedemann while supportive, an active and enthusiastic student of astronomy, was too caught up in the obligations of his church career to participate regularly in the mathematical inquiry where Nicolaus and Anna danced among the orbits of the planets about the sun.

Anna postulated three categories of data:

1. the sun observations by the Magellan expedition vs. the observations made from Seville (which differed by an entire day);
2. the full moon observations of the Magellan crew vs. those done at Seville (which matched) and
3. the inclination of the sun with respect to the seasons (the two sets of data matched as well as could be determined).

The conjecture Anna proposed was that only a Sun-centered system with revolving planets would satisfy the three data sets' triangulated intersection.

Copernicus had practically singlehandedly invented the idea of objective skepticism, one who wanted his own data and his own independent examination of the data offered by Anna. When more than a century later, the Royal Society was formed in England, their adopted motto was, and still is: Nullius in verba (Take no one's word for it).

Without a verified copy of the Magellan ships' logs, Copernicus had to frustratingly live with the reality that the copy made by Anna as being as close as he could get to the original source documents. In 1530, Frau Schilling and Father Copernicus set out to solve the *Mystery of the Missing Day*.

Without Cartesian coordinates (not coming until 1637), without logarithms (Napier, 1617) or a slide rule (Oughtred, 1621 to 1630), and without calculus (Leibniz and Newton in the 1670's), Anna and Nicolaus had to plow their way through prodigiously complex calculations with pen, paper, compass, ruler, straight edge, the Geometry of Euclid, and an abacus run by Anna.

Modern symbols for addition, subtraction, multiplication, and division, along with the equal sign were yet unknown in Europe. The introduction of symbolic algebra is decades away and Kepler. Fifty years after the death of Copernicus, mathematicians will still be using hexadecimal (base 60) numbers of the Babylonians invented some 4000 years earlier.

The paths of the planets had to be calculated entirely from geometry, the mathematics effectively unchanged since Euclid. One must wonder would this unlikely duo have embarked on their quest if they had known in advance how difficult this would become and how ill-timed new ideas were going to be in Europe in the mid-sixteenth century. Perhaps by 1530, both Anna Schilling and Nicolaus Copernicus were so emotionally invested in the Missing Day hunt that neither could abandon the pursuit. Maybe it was like the cliché of having a tiger by the tail: you do not have the ability to subdue the tiger, and you most assuredly do not dare let go—so you just continue to hang on.

This kind of collaborative effort among true intellectual equals bonds the participants in many ways more intimately than most married couples ever attain.

The relationship is not about physical intimacy but a much more delicate mental, emotional intimacy where your most personal thoughts are laid bare to intense, unflinching scrutiny. It requires knowing that your partner in this pursuit may be sharply critical but never hurtful, not even by accident, a special trust indeed. The partners involved in the pursuit of a passionately shared goal such as this effort understand each other at a depth and with a candor which people who do not work together may never reach.

The people of the general population in 1530 believed or at least pretended to believe, unquestioningly, what they were told to think by some combination of political and theological authority figures: politically, the local landowner, prince or king, and theologically priests and officials of the Catholic Church. For the most part, political authorities and theological authorities formed something of a cabal with little distinction between peasants and farmers (most of the population). Doubters were branded heretics, and heresy was a capital crime. The new reformers were proving themselves to be as brutal and repressive as the Catholic Church was at their worst. It was not a good time to be advocating radically new ideas unless you wanted to become "well known" in all the worst possible ways.

Copernicus had been reluctant to publish as much as anything, thwarting the urging and pleading of his friend Tiedemann Giese, because Copernicus felt his magnum opus was not yet sufficiently perfected, and it remained untested. Now Anna was providing the opportunity for a real test of much of what Copernicus had theorized.

The curia of Copernicus, just outside the walls of the Cathedral in Frauenburg, was not a mansion but certainly a comfortable domicile. However, even inside the castle walls, hidden away in his tower, Copernicus and Anna were not insulated from the tumult which raged outside, and often distressingly close at hand.

1531 to 1535 Making maps of Poland, Royal Prussia, the Ducal of Prussia, i.e., Europe from Lithuania to Paris must have been one of the more profitable enterprises as the religious allegiance of fiefdoms (states) and boundaries of Catholic vs Protestant territories changed almost daily. By 1535 Lutheranism had swept into much of northern Europe, taking over nearly all of what is now the German nation, Sweden, Denmark, and the Low Countries (Holland). Warmia was one of the few Catholic holdouts between the Rhine and the western edge of Russia. After being captivated by a couple of spectacular astronomical events in his college days, Copernicus came to astronomy largely by the way of books, the opinions of "natural philosophers," and the major figures of ancient academia, e.g., Aristotle, etc.

By contrast, Anna came to astronomy by the way of the docks of Danzig, listening to sailors sharing their experiences of navigating by the Sun, the Moon, and the stars. For her Astronomy was closer to cartography of the sky (star charts), mathematics and geometry all cooked in the caldron of application and studied from the deck of a ship at sea where those tools are used to survive.

By 1535 Anna had continued her pursuit of the Magellan Expedition Missing Day for the better part of seven years without pause to understand in greater detail and with more credible fundamental reasons, what we would now call physical science. As with any problem, Anna started by gathering as much information as she could find, starting all the way back to ancients, such as Aristarchus, and what few fragments could be found of their work. Since Anna had at first conjectured, she now wanted to test what evolved into her theory for the missing day: timekeeping by sidereal observations as distinct from solar ones. The moving earth model as articulated by Copernicus with our planet revolving about the sun had for Anna invited examination of several related phenomena for the sun, the moon and measured astronomical distances.

Through her trader merchant network and inquiring through Tiedemann Giese as early as 1531, Anna sought to find an expert astronomical instrument maker in Danzig. Several artisans were making their way to Danzig currently, particularly ex patriot Dutch watchmakers and metal workers, and Anna wanted to find someone able to build a more sophisticated sundial of her own design. Here the drawing skills, including constructed perspective that Anna had acquired in her travels to Florence, became most helpful in designing her new instruments and analyzing the resulting observations. A combination of good luck along with having a wealthy husband enabled Anna to study with teachers who had themselves been students of the great Brunelleschi (famed architect of the Santa Maria del Fiore Duomo in Florence – the largest free-standing masonry arch dome ever built, 1420 to 1436 and you can still visit it today).

A total lunar eclipse was observed on 15 June (1535), hardly a week before the Summer Solstice, which Anna used to invite Copernicus and Tiedemann Giese to an evening dinner on the July full moon (Wednesday the 23rd). Anna wanted to declare herself in league with Giese to convince Copernicus that he should publish his work.

Mathematics, astronomy, cartography, and navigation are all performing arts, one perfects their craft by the continuous practice of performing time and time again. To paraphrase a line from Shakespeare, "all art is but a brief and abstract chronicle of the times in which we live." The first 35 years of the 1500s were like no other time in recorded history. For centuries, the Catholic Church, in conjunction with the ruling aristocracies, had defined every aspect of social behavior, culture, and what passed as education. The Church decreed itself and its world as immutable and everlasting. Then suddenly, neither the world nor the Church was immutable, unchanging, or everlasting. Columbus's voyages started in 1492, and those who followed him found two new continents, North and South America. Then Martin Luther in Germany ignited a religious rebellion in 1517, which by 1535 had consumed more than half of Europe. Luther was followed by the Magellan Expedition voyage's return home (1522), which found a new ocean, the Pacific, the largest body of water on the planet.

If we knew so little about our world, even a few thousand miles from where we lived, then by extension, our knowledge of what was beyond the clouds must be minuscule if not almost non-existent. This view was punctuated by a meteorite that crashed in a wheat field just outside the walls of Ensisheim, Alsace, in November of 1492: the first time in recorded history that the impact of a meteorite had been observed, and it was in broad daylight. The literate faction of people in Europe must have felt as if they were living in the midst of a continuous earthquake.

There was no institution left standing that was not at least damaged if not in the process of total collapse.

People seldom, if ever, realize they are during a social and/or intellectual upheaval, a fact that only becomes apparent much later in retrospect. In 1535 Anna was living during the time when the entire cultural and intellectual fabric of Europe being shredded, and most importantly, it was being rewoven and rebuilt along entirely new lines. At this summer evening's small private dinner, Anna presented to her guests, Copernicus and Father Giese, a paradox for discussion. It seemed from Anna's perspective that the Church was fanatically concerned with what was written about the Sun, Moon, Earth, and the other planets. When in fact, the sky was arranged, however it was arranged, and it mattered not what was written about it.

If today a book is published which declares the Sun to be the center of the universe with all the planets revolving around it, the peasants and farmers – which is to say the vast center of the majority of the population – tomorrow will do just what they did yesterday, and every day before as this writing changes nothing about how we live daily. The universe stays the same.

Anna points out that there is a real revolution going on, which greatly affects every person's lives today, tomorrow, and for all the tomorrows to follow. "That revolution," Anna observes, "is not the Protestant Reformation or the Counter-Reformation nor what academics write about the sky.

The sweeping revolution taking place here, right around us, is being conducted by the craftsmen and tradespeople. They're changing the very way in which we live – and eat – and the Church completely ignores the entire process." Looking very perplexed, Tiedemann replies, "I'll admit to not seeing the sweeping revolution, or should I say, 'sleeping revolution' that you claim."

With an amused smile, Anna continues, "yes, 'sleeping revolution' it is, but just as we don't see the grass growing from hour to hour at the end of the summer, we realize the grass is shoulder high while in the spring it wasn't yet to our ankles. Recall the last time you traveled down the Vistula River from Torun to Danzig, or overland west from Danzig to say, Lubeck. How many windmills do you see that weren't there five or ten or fifteen years ago?"

Tiedemann, "They seem to be the Dutch 'smock' type, and now that you point to them, it does seem like they're springing up like weeds in a flower bed. But admittedly, I've not kept count."

"In that my father as well as my husband are both in the trading shipping business, they do count them, not directly but by noting how much the builders of these structures trade and ship. Most of those windmills are either grinding grain, wheat for the most part, or making iron. Much of the iron is used to make a faceplate for the moldboard of a plow, as well as parts for the plows. When taken along with the invention of the horse collar and the improved iron hardware that connects to the harness, there is a profound revolution afoot."

"Do continue," Copernicus says warmly, "I feel that I will be greatly edified as a result."

"The iron-faced plow pulled by horses allows a farmer to cultivate twice the land area in a third less time than before. By 'before', I mean within my lifetime. The iron also goes to making structural pieces for ships and a vast number of tools that previously were made of wood.

These changes are profound in how much food is available, where and at what price. Moreover, most windmills are being built just outside of the cities and towns, primarily to get out from under the guilds' control. Which means there are many more jobs available just outside of city walls than ever before."

Tiedemann, smiling indulgently, "I have the feeling you are leading us someplace."

"As a matter of fact, I am doing just so. Since the Church does not seem to notice what tradespeople and craftsmen do, perhaps because no one writes about it, I would like to suggest a game I call 'The Navigator's Challenge,' which I do want to write about. I got the idea when I was last in Torun on the western hills watching a couple of boats on the Vistula.

One boat was a Cog, heavily loaded and moving only slightly faster than the current. The other boat was a small sailing skiff, a shallop we see all the time, obviously a couple of boys out playing.

The small boat was essentially sailing a path parallel to the Cog and always going in the same direction, but speeding up and passing the

Cog then pulling ahead and to the side, the small boat would luff its sails and let the Cog pass it.

The small boat is making a winding path weaving along with the Cog. But to the passengers on the Cog deck, it appeared as if the small boat were traveling in circles around the Cog. I drew a little sketch of what it looked like from my vantage point, fixed on land high above the river on a hillside."

Explaining what she had observed, Anna had watched as the shallop stayed hard on the wind and sailing faster than the Cog. It pulled ahead but also gently steered towards the Cog. By the point where the shallop was leading the Cog, it was also dead ahead. The shallop continued this course, until it was abeam the Cog but now on its port side. Then the sailors on the Shallop lufted their sails so the Cog could catch up.

From the vantage point of the deck of the Cog it appeared as it the shallop had sailed a half-circle around the Cog, but the shallop had always headed in the same direction.

Anna paused to show her sketches.

ANNA'S STORY - 47

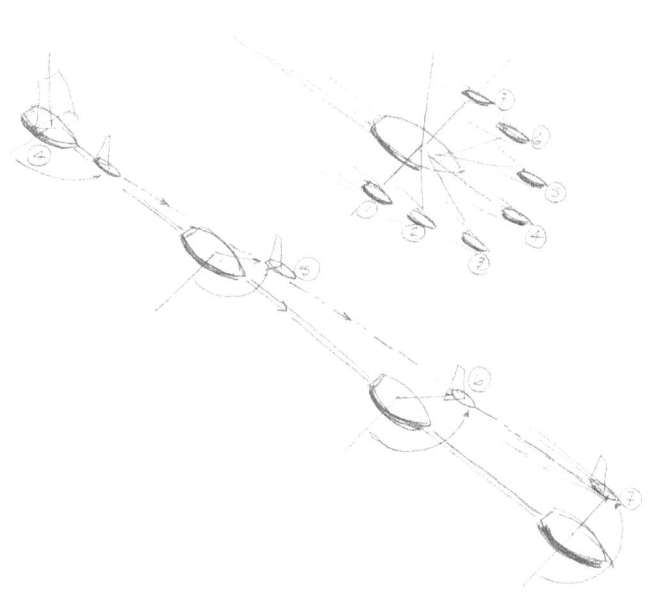

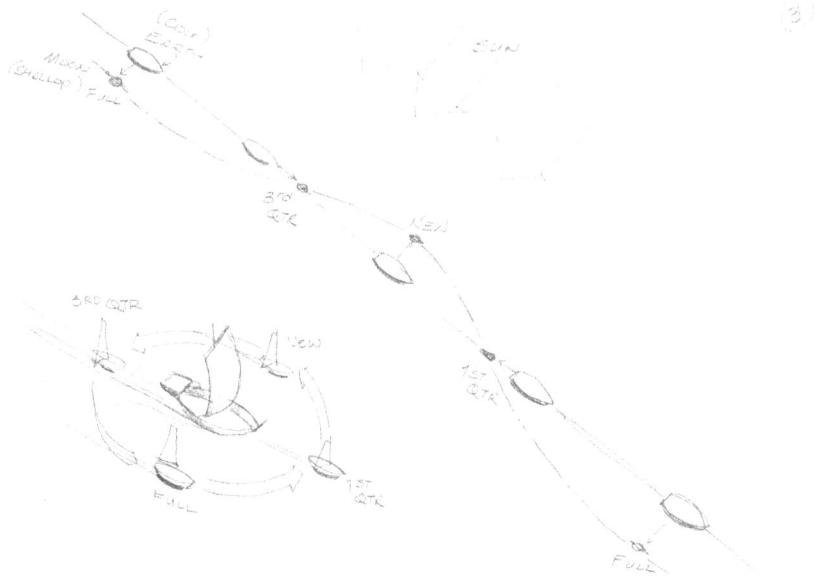

She was about to continue when Tiedemann slammed his hand down on the table – totally out of character as it was something he had ordinarily never done – and with wide eyes declared, "the path of the Moon about the Earth!

The Moon does not need to revolve about the Earth as the Moon parallels the Earth's orbit and weaves in and out of our path. Speeding up and slowing down in addition to weaving in and out of our orbit means we see the motion as revolving about us. Absolutely brilliant; as audacious as it is clever."

After a silence that felt as if it lasted for minutes, if not hours, Copernicus intoned with a sad, weary voice, "and not a person in the world will take us seriously. Moreover, if we were to advocate such an idea, then everything we write would be dismissed as the work of crazy people or fools or both."

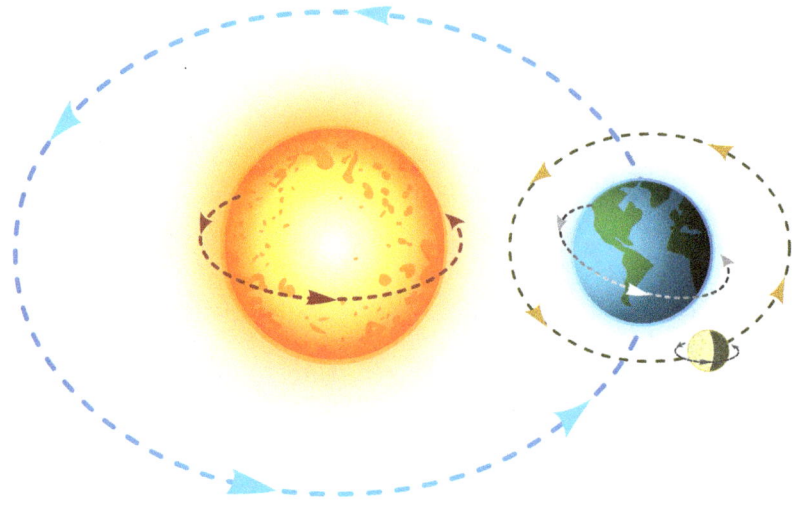

Sun - Earth - Moon Orbits

6

REVOLUTION.

"Ridiculous. I hate to admit that you are probably correct on this point as I would so very much like to pursue the idea."

This same issue arises more often with radical new science ideas than perhaps one might think, even in modern times, e.g., Alfred Wegener in 1925 proposed a theory of "continental drift," the direct precursor of plate tectonics. Wegener was completely shunned in geological circles, and his idea ignored until the 1960s, thirty years after Wegener died. Most famously, Newton did not use his own calculus to demonstrate the gravitational basis for planetary orbits – he used traditional geometry. As he explained in a letter to Edmond Halley, too few people understood calculus when Newton wrote "Principia Mathematica." He was so concerned that his argument about planetary orbits would get bogged down in explaining calculus. Newton decided to use traditional geometry instead. Moreover, Newton was convinced that the Sun moved through space by the same gravitational argument that explained the Earth revolving about the Sun. Newton was afraid that if he introduced a moving Sun, it would be considered such a crackpot idea no one would take the Earth argument seriously. One battle at a time, the audience can digest only a limited amount, and so Newton put the moving Sun idea off for another day (he never got around to that "other day"). Few people. other than those who have read Newton's letters to Halley, realize just how serious Newton was about a moving Sun – which we now know is entirely correct, as is plate tectonics.

The moon path Anna theorized, entirely correctly, was a serpentine route parallel to the Earth; the moon is seen to revolve about the Earth in a circular path only for Earth-bound observers.

The lunar revolution about the Earth takes approximately 27.3 days (Earth rotations) when observed against the stars, while the lunar cycle observed from Earth, full moon to full moon, is about 29.5 days. In 1536 Arendt van der Schilling, husband of Anna died. When Anna became a widow rather than "a married woman her own house in Frauenburg," Anna decided to take on the official role of "housekeeper" to Copernicus. This title included taking up some form of residence in his curia, a home less than 100 yards outside the walls of the Cathedral where Copernicus' tower is located. This was considered a perfectly legitimate arrangement in that Anna was a family relative, even if a distant one. However, in a village as small as Frauenburg, the arrangement, even though well within established rules, still set tongues to wagging.

The years 1537 and 1538 were a time in which an endless litany of mobs marching to and fro, conquest, pillage, and savagery with unknown thousands of victims in a boring list; boring unless you were the one doing the dying. The slaughter of innocent victims dwarfed even the wholesale massacres of the Crusades. There was more than enough blame for who was at fault in these murderous times to be shared by everyone on both sides.

By 1539, even with the tumult whirling about in the rest of the world, there were still those evenings when Nicolaus and Anna could retreat to Copernicus' tower and into their private world of planetary orbits and the mathematics of sidereal motion. In the last week of April that year, the curia of Copernicus was the setting for what was expected to be a quiet spring dinner and engaging conversation on the upcoming summer solstice event.

But on this very pleasant evening in late April their dinner was suddenly disrupted by a loud, frantic pounding at the door. Copernicus was astonished as his guests did not announce themselves in such a rude manner. Even more to his shock, when he opened the door, he found his good friend Tiedemann Giese standing there. Not waiting for the customary invitation Tiedemann all but barged into the house and walked directly to the table where Anna was still sitting. Copernicus found himself trailing along behind his friend.

Bishop Giese ignored the conventions of civil discourse, addressing both Anna and Nicolaus pronounced, "My children, this charming arrangement must stop, immediately, this very minute." Anna and Nicolaus were stunned. Not waiting for a response, Tiedemann continued. "A week ago, I learned that our Queen of Poland, Bona Sforza, had presided over the execution of an 80-year-old woman, Katarzyna Weiglowa, just seven days ago, 19 April, to be precise." Meeting uncomprehending stares, Giese explained that the woman had been imprisoned, presumably tortured, for a decade, ever since she was 70. Tiedemann remarked sadly, shaking his head, "burning her alive at the stake" may simply have been to appease the Counter-reformation zealots, but that is just the point. This whole Martin Luther business has gotten completely out of hand. The reformers and now the counter-reformers both seem to be intent on intimidating the other faction solely by means of increasing brutality. Any remaining shred of reason has long since been trampled and burned at the stake." Looking directly at Nicolaus, Giese added that the Bishop, Dantyszek (often Latinized to Dantiscus), had instituted formal proceedings against Copernicus about his "living arrangements," and explicitly accusing him of "concubinage. "Clearly, our bishop is trying to appease his superiors' furor over the continuing encroachment of Lutheranism into his own backyard by attacking even the smallest detail that might be viewed as contrary to Church doctrine."

A long silence ensued as Nicolaus and Anna looked at each other as if trying to decide what to say or do next. Tiedemann solved the indecision, "Anna, you must leave immediately. Tomorrow would not be too soon. I have neither the church connections necessary to dissuade

Bishop Dantyszek from what amounts to his witch hunt nor the political influence in Krakow to intercede with the Queen. But if you two continue here as you are now, then the future I see is that one or both of you will end up burned at the stake. Far better that you should live apart than die together, needlessly."

Looking only at Nicolaus, Anna spoke quietly, "Mikolaj, you have a great work which must be completed and then published. It has been the delight of my life to have participated in even a small way by asking questions and acting as your foil. I want more than anything to see your work in print. Do not squander your gift to posterity by flailing away at a battle we cannot possibly win.

It does seem that there is a great wave not of water rolling in from the sea but of fanatic passions sweeping over the whole of our world, so now with discretion being the better part of valor, I shall depart; return to Danzig and wait to see your life's passion in print.

With that, Anna stood, then walked wordlessly out of the room, gently touching Nicolaus' shoulder as she passed behind his chair. Finally, Tiedemann also arose, then, passing close to Nicolaus, said in a whisper, "only the three of us know I've made this visit, and absolutely no one else must ever hear of it." With those words as a farewell, Bishop Giese slipped away into the dark silence of the night.

Copernicus was now alone in a way he had not experienced in a decade: Anna, his collaborator and closest confidant, had been driven out; his neighbor, fellow canon, astronomer, and map maker, Scultetus (accused of secretly being a Lutheran), was fleeing from the same witch hunt and leaving town as quickly as he could, in addition to which his best friend Tiedemann was now posted far away and only able to visit on the rarest of occasions.

Anna left the following day at first light, accompanied by a single servant who also acted as a guard on the two days ride to Danzig. Within the week, two servants appeared in Frauenburg with a wagon to collect the items, books, furniture, astronomical instruments, clothes, and personal items which Anna had not taken with her.

Nicolaus continued to work at his calculations and manuscript, but it was completely impossible to ignore the emptiness of his tower and curia. For an entire decade, he and Anna had shared the toil and delights of grappling with one of the great enigmas of human history. They had marveled at how it was that only mathematics could illuminate the heretofore dark secrets of how the universe functioned. They had fed each other's souls and had found happiness.

With a gradual return to work, Copernicus began pondering the fact that no one remembers one martyr, much less the tens of thousands burned at the stake for what they felt were personal imperatives that were worth their life. However, Nicolaus began to see that he could use the very tool which had made it possible for Luther to engulf most of Europe in the conflagration that his rhetoric on theology had enflamed. Sitting at his desk with a copy of Euclid in his hands, Copernicus realized that the printing presses of every civilized city could become his mechanical army issuing forth entire legions on his behalf every month. One can only preach from one alter at a time, but all at once in countless places simultaneously, multitudes can hear your words with their eyes from the pages of your book. At the age of 66, Copernicus knew that far more of his career was behind him rather than ahead, and he must work more diligently than ever, even if he could not work physically as hard as he once was able to do.

At the end of the evening of Tiedemann's visit, Copernicus had been overcome with a rush of sensations he had not previously experienced. Tiedemann had been entirely correct. The whole Luther business was totally out of control and repeatedly sweeping back and forth across the land, wrecking what seemed to have been the civilized world. The world now plunged into a raging inferno of hatred. It left not only Copernicus feeling fragile, totally exposed, and vulnerable, but for the first time, Tiedemann began to feel that he was getting old – incredibly old and very tired.

Being immersed in a great, demanding project is a most effective tonic. As Copernicus started pouring more and more of his energies into his manuscript, he began to realize excellent therapeutic results.

7

PHILOSOPHY AND THEOLOGY

The full moon of May had not yet appeared when again Copernicus was interrupted at his evening supper by an insistent knocking. Concerned that it might be Tiedemann back again with more news of disaster, Copernicus hurried to the door. This time he was confronted by a stranger, from his attire clearly not a beggar and by his actions, not a robber. However, his clothing was not anything made in Warmia, his shoes were not any thing Copernicus had seen from anywhere in Prussia or Poland, and he appeared to be someone who had been riding for some while, days, perhaps weeks.

The unexpected guest spoke quickly, in German but with a western accent, trying to assure Copernicus of his best intentions and that he had traveled weeks to be a supplicant and student of the Great Astronomer.

Being rather wary of unannounced strangers, Copernicus asked, "traveled from where?"

The stranger introduced himself as Rheticus, born Georg Joachim de Porris, currently a mathematics professor at the University of Wittenberg. Utterly aghast, Copernicus demanded, "Wittenberg, the heart of Luther's inner sanctum, are you completely mad? You do know there is an official Church edict which bans all Lutherans from entering Warmia, or transporting any writing supporting the Luther cause into any part of Warmia? If you are found out, you will be fortunate to leave with your life, but more likely your last day here will be at the stake; and I might well end there too for not sounding the alarm that we have been invaded by Lutherans.

You have put yourself in grave danger just by being foolish enough to enter Warmia, and now have put both of us in even greater peril by appearing at my door. I have political enemies here who I'm certain report directly to the Bishop the identity of every person who enters or leaves my domicile

Rheticus had presents of three bound volumes, rather precious books to someone as learned as Copernicus. Them sight of the provided Copernicus with a flash of impromptu planning. "I can't have you and your traveling companion seen skulking out of here in the fading evening light. So you will put your horses in the barn and stay the night as if you have delivered the books I'd purchased from the last Frankfurt book fair. But then, as soon as it's light and you've had something to eat, you must both be on your way."

Seizing the interest Copernicus showed in the books, Rheticus continued to press his case by offering his commentary on how valuable the volumes would be as Copernicus worked to complete his magnum opus as promised in his anonymous letter.

"So you also have read my anonymous letter? How, where?" Copernicus asked.

"Johannes Schöner, after showing me your letter, made a copy himself for me when I visited his office at Nürnberg. You know he converted to Lutheranism and married?"

Rheticus smiled as he felt the mood between them was warming a bit. "No, Herr Doktor, you are not that well-read – yet – but you should be. However, the full book you promised in your anonymous letter would make you just that."

More than four decades later, Tycho Brahe gave the letter its enduring name: "de Commentariolus" or "The Little Commentary."

Schöner was famous as a mathematician, cartographer, astronomer and had his own printing business, including a full-sized printing press in his home – one of only a handful of mathematician astronomers who hugely impressed Copernicus. A few more questions, plus a combination of the books and recommendation by Schöner paved the way for a discussion that lasted through most of the night.

By morning Copernicus decided to make do with what he had available to him now and not waste time wishing for what he would like to have.

Thus the decision was made to allow Rheticus to stay in the guise of a cartographer from Padua. Since the voyages of Columbus, Da Gama, and Magellan, it would entirely reasonable that a cartographer would want to work on the mathematics of a new world map, particularly calculating the new latitudes and longitudes for correct projections onto a globe.

It was a watershed moment for Copernicus. By sheltering Rheticus, he became actively defiant of Church policy. So it was that in a moment, Copernicus decided that he would steer his life's ship on an entirely new course. For the first time in his life, Copernicus would act in direct contradiction to express edicts of the Church by conferring with a Lutheran, even if Rheticus was an astrologer and on balance, the deal seemed to represent a bargain of opportunity and convenience with someone that could be an agent of the devil. The final deciding factor may well have been the recognition by Copernicus that Rheticus would add one element not available to Copernicus anywhere else: the enthusiasm and the stamina of youth.

So, even if reluctantly, Rheticus could stay.

Ever since 1536 when Tiedemann had been made Bishop of Kulm (the modern name is now Chelmno), his visits to Frauenburg were necessarily less frequent than before. The one-way travel time from Kulm to Frauenburg was at best a three-day journey, two by boat on the Vistula and another hard day from Kulm to Frauenburg. However, Tiedemann used every required visit to the capital of Warmia on Church business to include a day or two in Frauenburg to visit with Copernicus.

After dinner in early 1540...

"It is a pleasure to see you again, Tiedemann. Thank you for taking the time to make this excursion to Frauenburg. Particularly this year, as I've heard that the spring floods on the Vistula are the worst in memory."

"The floods of water are merely an inconvenience compared to the floods of Lutherans, Anabaptists, Mennonites, along with the occasional Calvinist and Zwingli follower from Zurich followed by the washback of Catholics trying to recover the lost villages. The damage done by the various mobs on dry land under their Christian banners overwhelms anything that the river might do along its banks."

Sailor, "Nasty business, them damn Lutherans, eh, Bishop?"

Bishop Giese, "how can you tell, my son, all charred villages are covered by the same blanket of smoldering grief."

"One can escape the river by simply moving to higher ground. There is seemingly no haven from the fanatics. With all of the lands of the Teutonic Knights now firmly entrenched as the Ducal of Prussia and fanatically Lutheran, it is nearly impossible to travel overland directly east from Kulm to Allenstein." However, I've found the occasional ships' officer and the odd travel companion here and there to be most interesting."

"Aren't the accommodations on the riverboats rather primitive at their best? To say nothing of the crews and some of the travelers."

"Interesting thing that is, Nicolaus. The riverboats are an enjoyable mode of travel, and, yes, it is usually rather Spartan. However, I've found the occasional ships' officer and the odd travel companion here and there to be most interesting."

"How so, Tiedemann, or are you just trying to put an apron over an unpleasant condition you can't repair?"

"No, not at all, really. While we sit here pondering our whole universe's arrangement, on a scale we can't begin to comprehend, I find that the well-informed ship's officer is entirely refreshing. Their world is entirely accessible to direct inspection, a fantastic quilt of details. What is the best way to place cargo below and on the decks, who makes the best rope, who makes the best pulleys, where do you find the best charts and most current talk about currents and obstructions in the channels, who is landing what cargo on the docks at Danzig, what kinds of cargo are going downriver to the docks, what country has the best shipyards, where are the best cannons made – an endless list of particulars.

If you listen carefully, you can weave a very informative tapestry from this collection of bits. As we discuss the theory of astronomy, the crew of a boat practices the application of astronomy every hour. Their skill is often the difference between a shipwreck and a safe voyage.

But then there is the rare old traveler who does see a path emerging from what otherwise looks like a carpet of footprints on the beach just going hither and yon. They see a world where an increasing number of ships are being built to follow in the explorers' wake. The ships keep getting bigger, faster, and more heavily armed. They see what goods are increasing in trade volume and which are decreasing; also which ports are becoming busier and which are falling into disuse.

Nicolaus, you and all who study nature must be allowed to follow your observations, measurements, and mathematics where ever they take you. The old travelers are not constrained in the formation of their analysis by the pontifications of theologians, philosophers, or even the Church itself. Those who make pronouncements about how the natural world is organized based on nothing but an exalted view of their own ideas have no place in the study of nature, the natural world, and perhaps astronomy in particular.

Your problem only has one solution: how do the planets actually move. What is written about the Sun, Moon, Earth, and the planets does not change those objects. The challenge for us is to get the description right.

As you and I have discussed before, we need to get religion out of the pursuits of natural philosophy. Philosophy and theology should be in complete harmony with the physical world, with reality. As new information about our universe becomes available through astronomy, mathematics, cartography, and all measurement, it should be wholly embraced by philosophy and theology. As ancient Greeks stated with their word "physics," the study of nature, all such knowledge adds to the wonder and glory of this world. Instead, we've gotten into a position where theology and philosophy start with mere opinions.

If new measurements are in conflict with interpretation of the scriptures by theologians, then the theologians need to revisit their interpretations. The philosophers and theologians need to conform to reality rather than trying to hammer reality to fit their academic opinions.

"These things I say as a Bishop in the Catholic Church, so I am entirely familiar with our theology and philosophy."

"Ah, Tiedemann, ever the optimist. I don't think those who have power, authority, and consequently, great wealth will ever allow questions or investigations which might reduce their power or authority, or their attending wealth."

"Yes, Copernicus, I fear your skepticism is well-founded. However, just as in the course of time the technical details of your work will be augmented, perhaps eventually supplanting you entirely with more information and data from voyages such as those of Magellan who is sure to be followed by hundreds if not thousands of explorers. But then others may follow your lead just as future mariners will follow in the wake of Magellan. Consider Anna as your first acolyte, not blindly following your words but brilliantly following your example, she is pursuing her own, new view born of data and information. That, my dear colleague, my friend Nicolaus, Mikolaj, will be your great legacy – but only if you get your book into print.

Take advantage of what help this visiting mathematician, your 'cartographer,' can provide. And by the way, his Italian is atrocious. Please, do tutor him. Meanwhile, I will do everything I can to keep you from becoming a subject of interest that might warrant an inquiry by the Church."

Nicolaus and Tiedemann then raise their glasses in a toast ending the evening: "to the amazing beauty that is the harmony of the physical world – and the pleasure of finding things out."

In the span of hardly two decades, the entire political fabric of Europe had been shredded by the Protest Reformation movement ignited by Martin Luther. Prior to 1520, essentially all of Europe was Catholic. By 1540 at least half of the European area – Sweden, Denmark, Norway,

England, the German states from France to Russia and Lithuania in the east, and south to the Ottoman Empire– was controlled by Protestant governments.

8

EPILOGUE

The goal of Copernicus had always been to "fix" the Ptolemaic model. The spheres which Copernicus talks about in "The Revolution of the Spheres" are the crystalline spheres that carry the planets, not the planets themselves.

Copernicus does do by example to make a case for objective skepticism being allowed, protected in fact, to ensure that new, better ideas based on observation and data are developed and embraced. Rational discourse about data should direct our way forward, not fanatical protection of entrenched philosophies, mere opinions, which often find themselves at odds with reality.

Young Rheticus enables Copernicus to finish the manuscript for his book and carries those manuscript pages to a Nuremberg printer. When the first-ever bound copy is delivered directly from the printer and handed to Copernicus in his sickbed, he embraces the volume and takes his last breath. Or at least that is the favored biographical version.

Sadly, after Anna Schilling left Frauenburg in 1539, no verifiable document confirms where she went next. Many stories persist that Copernicus visited Anna in Danzig after she left Frauenburg, but there are no documents to support such claims. There are some surviving letters that relate to third-hand reports saying Anna was seen in Frauenburg canon, but it is also rumored that when she tried to return to Frauenburg to sell her house, her request to enter Frauenburg was denied. She sold the property through a third-party agent.

In the 1990s a house in Gdansk (called Danzig in the time of Copernicus) was being renovated, and brick masons working in the basement found a fragment of a wooden chest marked "Anna S 1539 AD." The house is now a Copernicus tourist attraction in Dansk. There, what little trail there was – ends.

Somehow an "urban legend" attached itself to Copernicus' work as "the book nobody read." A delightful volume by Owen Gingerich, "The Book Nobody Read," decimates that bit of nonsense. Gingerich tracked down more than 600 first and second-edition copies of "The Revolutions" over the course of thirty years to examine the marginalia in those early editions. From the annotations found by Gingerich, many if not most of those owners did indeed read Copernicus' book and in careful detail. Gingerich found the copies owned by Galileo, Kepler, Brahe, Huygens, Newton, Leibniz, and Edmund Halley, among the hundreds he was able to see and photograph. A 2nd edition copy heavily annotated, once owned by Edwin Hubble, is on display at the Huntington Library in San Marino, CA; I was able to view the book in the course of gathering information for this little booklet. If you can find a first edition of "On the Revolution of the Heavenly Spheres" for sale, be prepared to bid well in excess of $2 million dollars for it.

In 1616 the Catholic Church put the works of Copernicus on the Index Librorum Prohitorum (list of forbidden books), a ban that was not lifted until 1758.

It will take almost another century after Copernicus before Kepler (b 1571 – d 1630) can analyze the new and better data of Tycho Brahe and show by mathematics that the planets move in elliptical orbits about the sun (Kepler published his "Laws of Planetary Motion" between 1609 and 1619). Thus, he quietly destroyed the "perfect spheres" containing the planets and the paths of "perfect circles," which had confined the planets and ruled astronomy by opinion alone for more than a thousand years. The Brahe data of the 1577 comet also wrecked the crystalline spheres as his data made clear the comet came from into our solar system, passing closely by Saturn then around the Sun, passing close to Earth (closer than Mars), and then headed back into deep space (source: JPL C/1577 V1).

When the Royal Society formed in England (28 November 1660), their motto was and still is: "nullius in verba," or in English, "on the word of no one;" which has also been translated as "take nothing on authority." Now 460 some years after Copernicus, and 356 years after the Royal Society's formation, it seems particularly odd and monumentally tragic that education in the US is rapidly regressing to where students are expected to "learn" by accepting everything on authority of the teacher and/or textbook. Learning and understanding by direct experience of one's own investigations cannot be overvalued. While we have more information available to us, more quickly accessible than any time in history, we must remember that data and a collection of facts do not constitute understanding, much less do they substitute for wisdom.

9

Science, Technology, Engineering and Math

Science is about real, tangible things that move, fly, explode, burn, haul loads, float, record images, stop diseases, and explore the universe. It includes all the interesting, exciting "stuff" of the times in which we live.

The language all scientists and other STEM professionals use is mathematics. Mathematics is the study of patterns. With math modeling available to the general public through modern technologies like graphing calculators, anyone can learn to speak this language.

ABOUT THE AUTHOR

Engineer, artist, educator, and former race car driver, resides with family in Portland, Oregon in the St Johns Neighborhood. *Anna's Story: In Pursuit of the Mysterious Missing Day* is the first historical novel in the Scientific Light and Illustration series. Jeffery shares his lifelong passion for engineering and art through workshops for children of all ages. Project workbooks will accompany these novels for readers, teachers, and students inspired to become modern tinkers, makers, and innovators.

www.ingramcontent.com/pod-product-compliance
Lightning Source LLC
Chambersburg PA
CBHW070045230426
43661CB00005B/761